"Cutting Edge" is circulated by the Kentucky Arts Commission and the Southern Arts Federation, funded in part by the National Endowment for the Arts and the state arts agencies of Alabama, Florida, Georgia, Kentucky, Louisiana, Mississippi, North Carolina, South Carolina, Tennessee, and Virginia. Additional funding was provided by the Louisville Art Gallery, The Pace Trust, Humana, Inc., and the Atlantic Richfield Foundation on behalf of Anaconda Aluminum Company.

Library of Congress Catalogue Card Number 81-620014
ISBN 0-939058-01-4

Design: Images, Julius Friedman, Louisville, Kentucky
Photography: Warren Lynch, Louisville, Kentucky

CUTTING EDGE

This catalogue documents an exhibition of work in metal organized to tour the southeastern United States through the Kentucky Arts Commission and the Southern Arts Federation in 1981/82. The prospectus encouraged a broad interpretation of the theme, the "cutting edge." Work by artists and craftspeople nationwide, produced between 1978 and 1981, was selected by three jurors with different points of view: a curator with a strong historical perspective, a traditional knifemaker, and a contemporary goldsmith. Each offered names of artists and craftspeople to be invited and each came to Kentucky to jury, independently, work submitted in an open competition. The result is a collection which primarily focuses on functional objects—knives, daggers, cookie cutters and even a sled—but also includes small sculpture as well. Metal is the dominant medium—steel, iron, bronze and brass among them—with subsidiary materials essential to the object, including leather, wood, ivory, bone, and bamboo.

This is a small show of fine craftsmanship, sound design, and subtle spirit. It is not a flashy show. Its strength is not obvious from superficial or hasty viewing. The humor, life, and expression of the objects surface with close scrutiny.

The concept for "Cutting Edge" was sparked during a conversation between goldsmith Jim Cotter from Colorado and Kentucky Arts Commission Traveling Exhibition Service director Albert Sperath, whose own work shows a proclivity for sharp things. Their idea was to cross the boundaries of art and craft through a familiar theme.

The Kentucky Arts Commission would like to thank Jim Cotter, Julius Friedman, Warren Lynch, and Linda Anderson for their insight. We are also truly grateful to the Louisville Art Gallery and its director, Roberta Williams, who agreed to organize the exhibition cooperatively with Albert Sperath. This show and catalogue would not have existed without the help of other unnamed people and the organizations that gave so generously to the success of "Cutting Edge."

We would also like to thank jurors Robert Ebendorf, Daryl Meier, and Richard J. Wattenmaker, who approached their work with an excitement and confidence that was inspiring.

Anne Ogden, Director
Arts Programs
Kentucky Arts Commission

When Albert Sperath came to me two years ago with the idea of doing a thematic show on cutting edges, I was delighted. I thought that it would be a superb venture. It fit in well with a series of exhibitions the Gallery has done which focus on everyday objects as art forms. I think this is an important idea to be addressed. This kind of theme helps to reinforce the idea that good design is an important part of everyday life and that aesthetic considerations need not be separated from utilitarian functions. The work of the artists included in this exhibition reflects this philosophy.

Roberta Williams, Director
Louisville Art Gallery

JURORS' STATEMENTS

Richard J. Wattenmaker

The self-discipline of maintaining contact with contemporary crafts is important to me. The paucity of informed criticism devoted to this ever-expanding segment of the art world is an obstacle. This can be overcome by participating in exhibitions such as "Cutting Edge," wherein artist, critic, collector and public alike can enlarge their appreciation, thereby enhancing both production and perception. The treasured wooden, glass, metal, fibre and pottery creations from the past that we preserve are the tangible legacy of artisans. They constitute the living tradition of craftsmanship descended in an unbroken chain from primordial times. Although the term revival has been used, perhaps renewal is better suited to this multiform contemporary activity.

In making my nominations and selections I relied upon what I knew had been achieved by several of the outstanding artist-blacksmiths working today. In considering entries I sought a directness of conception, skilled execution, a sense of the limits of the medium and an alertness to physical and conceptual proportion. Although the composite selection is not the result of one mind, I hope that the cross-section here offered will fulfill its purpose, stimulate curiosity and provide enjoyment to the many visitors who will see this exhibition.

Robert Ebendorf

I feel that the time has come when we can look for an integration of concept, materials and process—an integration that is beyond the decorative and beyond the craft. I felt justified in holding this expectation in judging the pieces.

I have looked for and found pieces that act as metaphor, transporting one by means of the medium to image and idea. These pieces go beyond technical skills and expertise. Many have come of age artistically and begin to meet man's historic need for identity and sustenance.

In instances where I have rejected a piece I often sensed a lack—not in skill, but in taking the responsibility to find and visually realize an idea.

I recognize my fallibility.

As a fellow craftsperson I felt pride in the level of the entries.

Daryl Meier

I used much the same criteria for selecting the invitational and competitive segments for "Cutting Edge"—craftsmanship, design concept, clean logical lines and honesty in the historical context of the work whether ancient, recent, contemporary or avant garde.

Phillip Baldwin
Portland, Oregon
"Going and Coming #2"
metal, wood
75 cm. L x 22.5 cm. W x 3.5 cm. H

INVITATIONAL SECTION

David Campbell
Eugene, Oregon
Letter Opener
mild steel
20.5 cm. L x 2.3 cm. W x 1.9 cm. H

Kenneth M. Coleman
Richmond, Virginia
"Knife For An Evening Out"
440 c stainless steel, rosewood,
sterling silver, copper, bronze
11.7 cm. L x 2.5 cm. W x .8 cm. H

Glen Gardner
Burnsville, North Carolina
"Claw Knife"
steel, bronze
7.5 cm. L x 2.9 cm. W x 1.7 cm. H

Robert A. Griffith
Montrose, Pennsylvania
"Assault On A Cube," Form #2
steel, brass, Plexiglas, paint, curly maple
15 cm. L x 7.5 cm. W x 10 cm. H

Louis Brent Kington
Makanda, Illinois
Anvil Cookie Cutter
iron, copper
8.2 cm. L x 10 cm. W x 8.5 cm. H

Louis G. Mills (Yasutomo)
Wyandotte, Michigan
Japanese Tanto (dagger)
Damascus steel
30 cm. L x 2.9 cm. W x .8 cm. H

Gary Noffke
Farmington, Georgia
Knife (Shark)
steel, bone
27.5 cm. L x 6 cm. W x 3.5 cm. H

Will Pass
Merritt Island, Florida
Arkansas Toothpick
stainless steel, nickel silver, ivory
30 cm. L x 12 cm. W x 2.5 cm. H

PASS

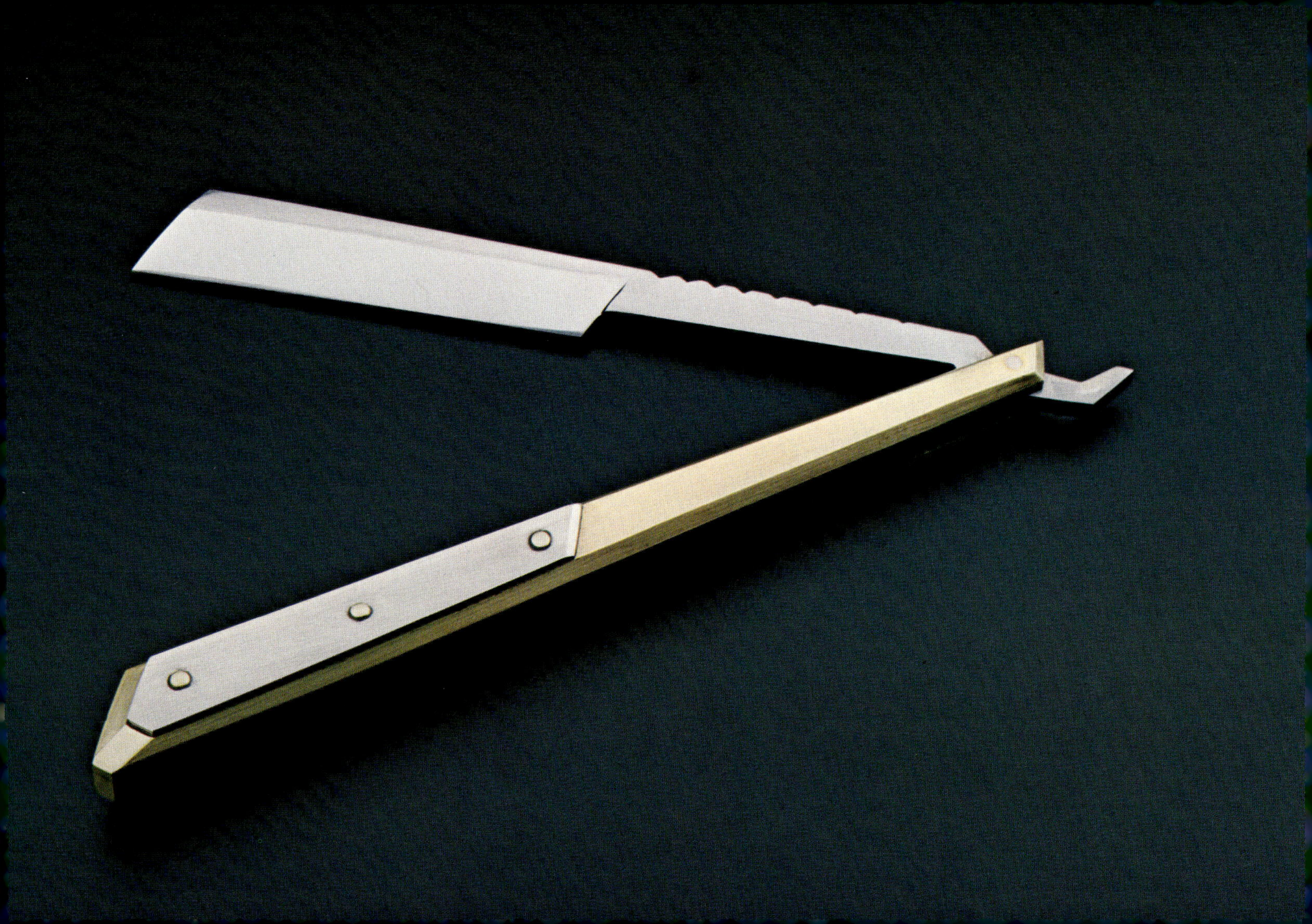

David Rogers
Coleman Falls, Virginia
Straight Razor
brass, nickel silver, tool steel
14.5 cm. L x 2.2 cm. W x 1.1 cm. H

Joel A. Schwartz
Deansboro, New York
Cheese Cutter
steel
20 cm. L x 11.5 cm. W x 2 cm. H

Robert E. Timberlake
Timberlake, North Carolina
"Aikuchi"
Damascus steel, birdseye maple, sterling silver
30 cm. L x 3 cm. W x 2 cm. H

Jim Wallace
Memphis, Tennessee
Sled, #7
iron, cherry, walnut
102.5 cm. L x 33.2 cm. W x 24.4 cm. H

Bob Coogan
Smithville, Tennessee
Garlic Knife
silver, brass, d2 steel, mahogany
15.5 cm. L x 3 cm. W x 4.5 cm. H

JURIED SECTION

Michael Croft
Tucson, Arizona
"Fenestrator"
01 steel, brass, corian
28.5 cm. L x 5 cm. W x 1.5 cm. H

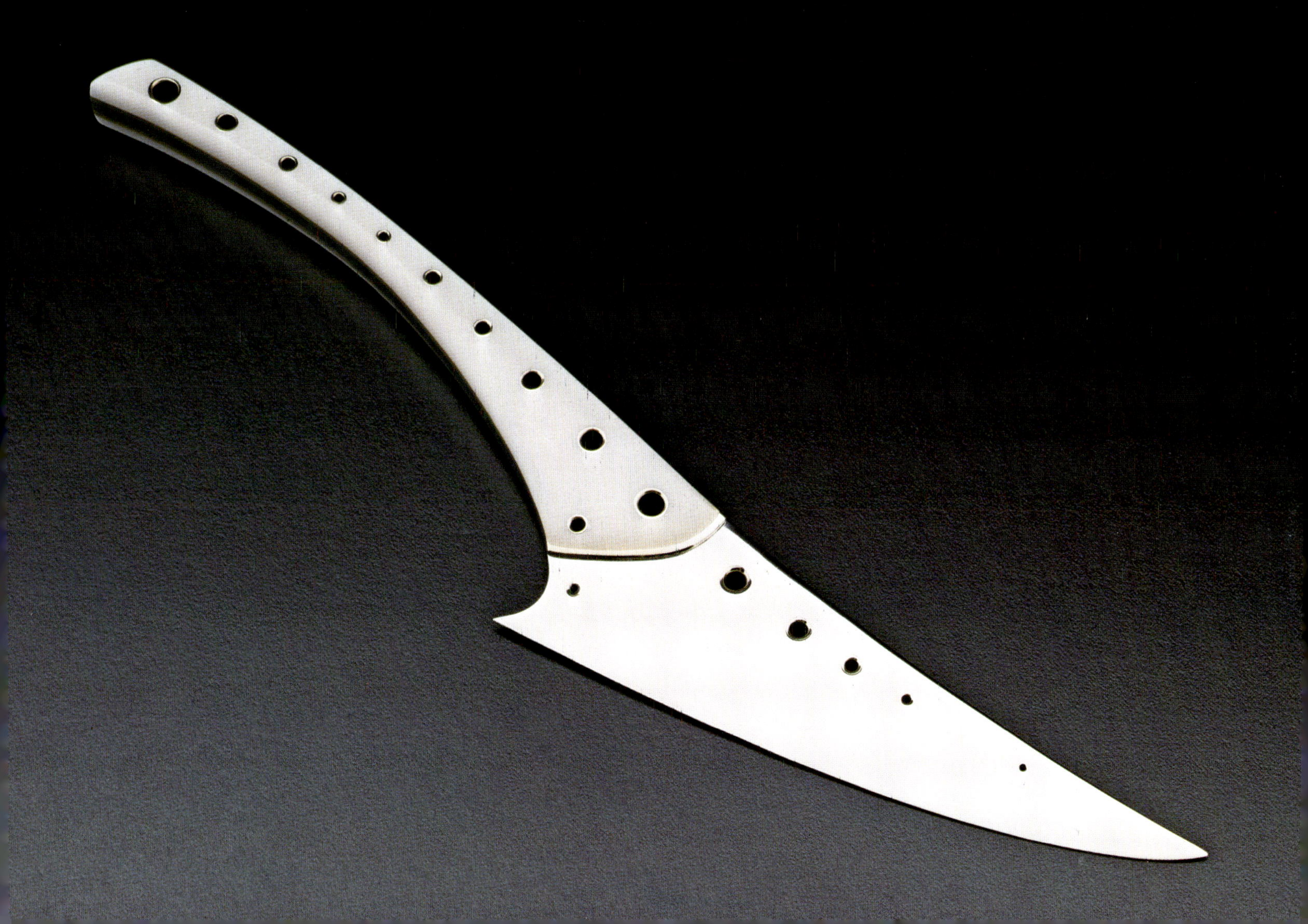

Mark Diamond
Houston, Texas
"Blade"
steel, silver, gold, enamel, ebony
24.5 cm. L x 11.5 cm. W x 10.4 cm. H

Scott C. Fredenburg
Carbondale, Illinois
"Wind Worshiper"
01 steel, brass
28 cm. L x 3 cm. W x 1.2 cm. H

KHALSA
KIRPAN

Jot Singh Khalsa
Boston, Massachusetts
Utility/Caping Knife (small hunting knife)
cocobolo wood, brass bolts, stainless steel
14 cm. L x 2.5 cm. W x 1 cm. H

Mark Knuth
Ames, Iowa
Skinner Knife
tool steel, Damascus, delrin
10.2 cm. L x 3.5 cm. W x 1.5 cm. H

Jan Brooks Loyd
Newell, North Carolina
Spatula (egg)
steel, brass, walnut
38.2 cm. L x 10 cm. W x 5.6 cm. H

Dan Maragni
Engraving—Marcello Pedini
Northport, New York
Borgia Dagger (Cesare)
Damascus steel, nickel silver, wengay
27 cm. L x 6 cm. W x 2 cm. H

Bruce Metcalf
Kent, Ohio
"Stabbed Moon, Scarred Ground, and Knife Sky"
metal, plastic, paint
25 cm. L x 25 cm. W x 2.5 cm. H

Morris C. Pulliam
Shelbyville, Kentucky
2-Blade Pocket Knife with File Work Springs
01 tool steel, nickel silver, bone
7.5 cm. L x 2.5 cm. W x 1.25 cm. H

PULLIAM
PULLIAM

Lynda Ross
Missoula, Montana
Fish Knife
mixed
16.5 cm. L x 3 cm. W x 1.6 cm. H

Richard Sexstone
Layton, New Jersey
Paring Chisel
tool steel/mild steel, lignum vitae
32.5 cm. L x 2.5 cm. W x 2.5 cm. H

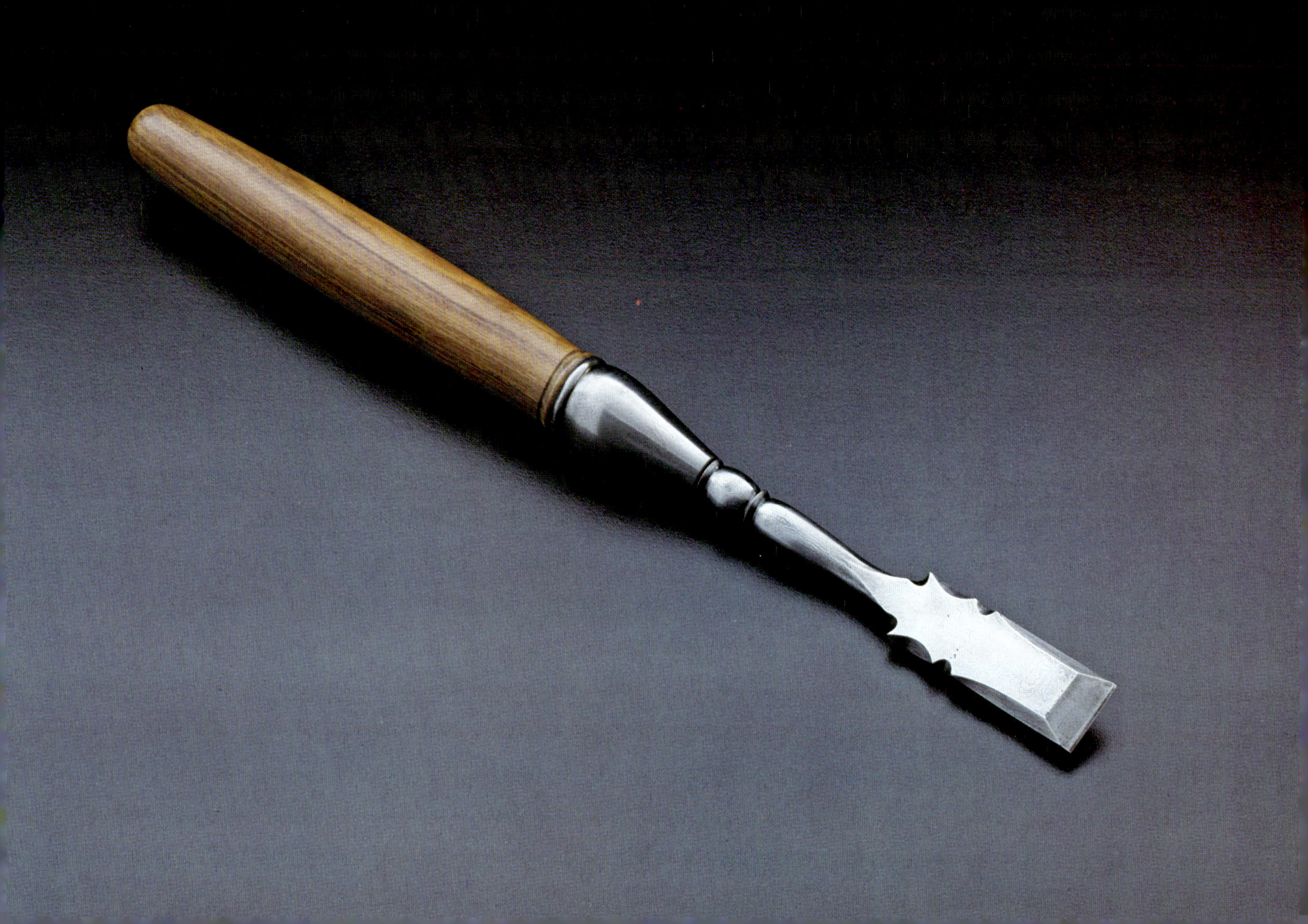

Stanley Winkler
Murphysboro, Illinois
Axe
Damascus steel, hickory
52 cm. L x 19 cm. W x 2.6 cm. H

John D. Wittersheim
Adrian, Michigan
Stiletto/Sheath
steel, brass, leather
42.5 cm. L x 7.5 cm. W x 5 cm. H

INVITATIONAL SECTION
Not photographed in catalog

Phillip Baldwin
Portland, Oregon
"Gallery Opener #2"
metal, wood
29 cm. L x 4 cm. W x 2 cm. H

David Campbell
Eugene, Oregon
Scissors
mild steel, high-carbon steel
19 cm. L x 10 cm. W x 1 cm. H

Kenneth M. Coleman
Richmond, Virginia
"Knife For A Sunday Outing"
440 c stainless steel, ebony,
sterling silver, copper, bronze
11.8 cm. L x 2.4 cm. W x .9 cm. H

Glen Gardner
Burnsville, North Carolina
"Goose Knife"
steel, bronze
5.5 cm. L x 5.5 cm. W x .6 cm. H

Robert A. Griffith
Montrose, Pennsylvania
"Cutting Edge," Form #1
steel, Plexiglas, brass, curly maple
21 cm. L x 7.5 cm. W x 15 cm. H

Louis Brent Kington
Makanda, Illinois
Hammer Cookie Cutter
iron, copper
13.5 cm. L x 8 cm. W x 5.5 cm. H

Louis G. Mills (Yasutomo)
Wyandotte, Michigan
Japanese Tanto (dagger)
Damascus steel
36 cm. L x 2.9 cm. W x .9 cm. H

Gary Noffke
Farmington, Georgia
Knife (Bamboo)
steel, bamboo
22.5 cm. L x 2.25 cm. W x 1.2 cm. H

Robert E. Timberlake
Timberlake, North Carolina
"Khanjar"
Damascus steel, rosewood, sterling silver
24 cm. L x 3 cm. W x 2.5 cm. H

Jim Wallace
Memphis, Tennessee
Axe
wrought iron, steel, wood
86.25 cm. L x 25.6 cm. W x 3.1 cm. H

JURIED SECTION
Not photographed in catalog

Bob Coogan
Smithville, Tennessee
Folding Hunter, I
brass, d2 steel, wood
15.5 cm. L x 4 cm. W x 1.25 cm. H

Jan Brooks Loyd,
Newell, North Carolina
Spatula (serving)
steel, brass, walnut
35 cm. L x 14.5 cm. W x 6 cm. H

Dan Maragni
Northport, New York
"New York Toothpick"
Damascus steel, bronze, copper, padouk
35 cm. L x 5.5 cm. W x 3 cm. H

John D. Wittersheim
Adrian, Michigan
"Sword/Mace Form"
steel, brass, walnut
43 cm. L x 13.5 cm. W x 18 cm. H